I0475781

TABLE OF CONTENTS

LIST OF TABLES

FISCAL YEAR 2014 REPORT TO CONGRESS ON FEDERAL CLIMATE CHANGE EXPENDITURES

1. INTRODUCTION

> *"We can't have an energy strategy for the last century that traps us in the past. We need an energy strategy for the future – an all-of-the-above strategy for the 21st century that develops every source of American-made energy."*
>
> —President Barack Obama, March 15, 2012

> *"We will continue to lead by the power of our example, because that's what the United States of America has always done. I am convinced this is the fight America can, and will, lead in the 21st century. And I'm convinced this is a fight that America must lead. But it will require all of us to do our part. We'll need scientists to design new fuels, and we'll need farmers to grow new fuels. We'll need engineers to devise new technologies, and we'll need businesses to make and sell those technologies. We'll need workers to operate assembly lines that hum with high-tech, zero-carbon components, but we'll also need builders to hammer into place the foundations for a new clean energy era."*
>
> —President Barack Obama, June 25, 2013

The following is an accounting of Federal funding for climate change programs and activities, both domestic and international, included in the fiscal year (FY) 2014 President's Budget. This report is provided in response to Title IV, Division E, Section 425, of P.L. 112-74, the Consolidated Appropriations Act of 2012 continued under P.L. 113-6, Consolidated and Further Continuing Appropriations Act, 2013:

> *Not later than 120 days after the date on which the President's fiscal year 2013 budget request is submitted to Congress, the President shall submit a comprehensive report to the Committee on Appropriations of the House of Representatives and the Committee on Appropriations of the Senate describing in detail all Federal agency funding, domestic and international, for climate change programs, projects and activities in fiscal year 2011, including an accounting of funding by agency with each agency identifying climate change programs, projects and activities and associated costs by line item as presented in the President's Budget Appendix, and including citations and linkages where practicable to each strategic plan that is driving funding within each climate change program, project and activity listed in the report.*

1.1 BACKGROUND

The U.S. Government's portfolio of climate change programs and cross-cutting initiatives focuses on advancing our understanding of climate change and its impact on our communities; advancing the development and introduction of energy-efficient, renewable, and other low- or non-emitting technologies; improving standards for measuring and registering emissions reductions and supporting preparedness and resilience to climate change impacts. Many elements of the Administration's climate change portfolio are designed to provide incentives for greenhouse gas (GHG) emissions reductions domestically to support community-based preparedness and resilience efforts, to ensure that Federal operations and facilities continue to protect and serve citizens in a changing climate, and to promote international initiatives focused on concrete actions toward reducing greenhouse gas emission and enhance climate preparedness globally. The Obama Administration has set a U.S. GHG emissions reduction target in the range of 17 percent below 2005 levels by 2020 and approximately 83 percent below 2005 levels by 2050.

Climate and Global Change Research and Education. Through the U.S. Global Change Research Program (USGCRP), U.S. scientists are conducting world-class research on climate and global change. The USGCRP coordinates scientific research across 13 Federal departments and agencies with the mission of "build[ing] a knowledge base that informs human responses to climate and global change through coordinated and integrated Federal programs of research, education, communication, and decision support."[1]

Reducing Emissions through Clean Energy Investments and Standards. The Administration is pursuing a wide range of initiatives that reduce greenhouse gas emissions through clean energy technologies and policies. The Administration has made the largest clean energy investment in American history and these investments have allowed the U.S. to double America's renewable power generation since 2008.

International Leadership. Under President Obama's leadership, the United States has engaged the international community to promote sustainable economic growth and to meet the climate change challenge through a number of important venues including: international climate negotiations in Copenhagen (2009), Cancun (2010), and Durban (2011); the Major Economies Forum, the Clean Energy Ministerial, the Climate and Clean Air Coalition, and the Asia-Pacific Economic Cooperation (APEC) Summit.

Climate Change Adaptation. At the request of President Obama, an interagency Climate Change Adaptation Task Force has crafted recommendations for how Federal agency policies and programs can better prepare the United States to address the risks associated with a changing climate. Federal agencies have released their first-ever climate change adaptation plans to help ensure smart decisions that protect our investments and safeguard the health and security of our communities, economies, natural resources, and infrastructure from the impacts of severe weather, rising sea levels, and other changing climate conditions. The Task Force has also helped develop the National Fish Wildlife and Plants Climate Adaptation Strategy to guide ecosystem adaptation and resiliency efforts.[2]

---

[1] http://www.globalchange.gov/about
[2] www.wildlifeadaptationstrategy.gov

The budget information presented in this report reflects the Administration's commitment to address climate change while preserving a strong American economy. The President's 2014 Budget proposes over $21.4 billion for climate change activities. This amount is $1.2 billion, or 5 percent, lower than the 2013 enacted level for climate change programs, activities, and related tax policies.

1.2 REPORT OUTLINE

The President's 2014 Budget supports a wide range of climate change-related research, development, and deployment programs, voluntary partnerships, and international aid efforts. This report presents the expenditures associated with this portfolio of activities in five main categories – science, technology, international assistance, tax provisions, and adaptation efforts associated with natural resource adaptation – as described below:

- **Climate Change Science.** This category encompasses the U.S. Global Change Research Program (USGCRP).

- **Clean Energy Technology.** Clean Energy Technology incorporates a variety of technology research, development, and deployment activities – including voluntary partnerships and grant programs – that support reductions in greenhouse gas emissions and reliance on fossil fuels. This category comprises work on clean energy systems and sources such as geothermal, solar, wind, biomass, nuclear, and emerging sources such as water power. It also includes programs or technologies or practices that help improve energy efficiency or reduce energy consumption, such as building efficiency, more effective transmission or distribution of electricity, and vehicle technologies that improve engine efficiency or fuel economy.

- **International Assistance.** This category describes elements of a "whole of government" approach to mobilize a wide range of resources and make use of bilateral and multilateral assistance tools. The core budget includes resources for a coordinated set of programs designed to ensure an effective balance across the three pillars of the global climate effort: Adaptation, Clean Energy, and Sustainable Landscapes.

- **Energy Tax Provisions.** This category includes tax incentives for investments in certain energy technologies, and energy payments that can be used in lieu of certain tax credits. These incentives promote deployment of energy efficient or alternative energy technologies, which may help reduce greenhouse gas emissions.

- **Climate Change Adaptation, Preparedness, and Resilience.** There are numerous efforts across the Federal Government for preparing and building resilience to the impacts of climate change on various critical sectors, institutions, and agency mission responsibilities. This concept is also known as "adaptation." Led by the Interagency Climate Change Adaptation Task Force, and using risk management principles, agencies are working to ensure they can continue to perform their missions in the face of climate change. Successful preparedness efforts often involve integrating climate change considerations into existing agency programs, projects, and activities rather than establishing separate and distinct programs. This creates a challenge when attempting to fully account for all adaptation resources. While the Administration continues to develop methodologies to account for a broader suite of adaptation programs across all critical sectors, an interim category, described further in section 6, summarizes certain activities at the Department of the Interior designed to promote preparedness and resilience. The activities at the Department of the Interior reflect interagency efforts to address key adaptation challenges that cut across the jurisdictions and missions of individual Federal agencies, and affect fresh water, oceans and coasts, and fish, wildlife and plants.

The following sections provide further detail in each of these five areas.

Table 1

Summary of Federal Climate Change Expenditures

(budget authority in millions of dollars)

Summary of Climate Expenditures[1]	FY 2012 Enacted Budget Authority	FY 2013 Enacted Budget Authority	FY 2013 Current Budget Authority[8]	FY 2014 Proposed Budget Authority	Change in Budget Authority 2013-2014
US Global Change Research Program (USGCRP)	2,506	2,509	2,463	2,658	+149
Clean Energy Technologies	6,121	6,088	5,783	7,933	+1,845
International Assistance [2,7]	958	851	797	893	+42
Natural Resources Adaptation	88	95	95	110	+15
Energy Tax Provisions That May Reduce Greenhouse Gases [3,4]	5,052	4,999	4,999	5,129	+130
Energy Payments in Lieu of Tax Provisions [5,6]	5,080	8,080	8,080	4,710	-3,370
Adjustments for programs included in multiple categories	*-24*	*-24*	*-22*	*-23*	---
Total [1,7]	19,781	22,598	22,195	21,408	-1,189

Footnotes:

[1] Budget Authority provided in millions of dollars and are current as of June 21, 2013. Discrepancies with other published documents may result from rounding and improved estimates.

[2] International Assistance includes congressionally appropriated assistance by core agencies (i.e. Department of State, Department of Treasury, US Agency for International Development) as well as complementary agencies (e.g., Environmental Protection Agency), but does not include indirect climate assistance nor development finance and export credit agencies.

[3] Tax incentives related to climate change included in this report were projected at about $23.5 billion over five years (2014-2018). These estimates do not reflect the extension of several temporary tax provisions by the American Taxpayer Relief Act of 2012.

[4] Tax expenditures are estimates of the revenue losses due to a tax preference. While not exactly equivalent to budget authority, tax expenditure estimates are included for completeness.

[5] Firms can take an energy payment in lieu of certain tax credits. The payments are considered outlays and are direct substitutes for the energy tax provisions. Estimates have been included in all columns for completeness.

[6] Energy payments in lieu of tax credits included in this report are currently projected at $9.1 billion over five years (2014-2018).

[7] The International Assistance total contains funds that are also counted in the USGCRP and Clean Energy Technology totals. Table total line excludes this double-count.

[8] Current Budget Authority for FY 2013 throughout this document reflects the amount the program has available for the year calculated as the appropriated amount (as reported in the FY 2013 Enacted column) minus the reductions pursuant to the Budget Control Act of 2011 (P.L. 112-25) sequestration order issued on March 1, 2013, and accounting for any known and applicable reprogrammings, transfers, or other related adjustments. Estimates are current as of June 21, 2013 and are subject to change.

2. CLIMATE CHANGE SCIENCE

The U.S. Global Change Research Program (USGCRP) was mandated by Congress in the Global Change Research Act of 1990 (P.L. 101-606) to improve understanding of uncertainties in climate science, including the cumulative effects on the environment of human activities and natural processes, develop science-based resources to support policymaking and resource management, and communicate findings broadly among scientific and stakeholder communities. Thirteen departments and agencies participate in the USGCRP. The Office of Science and Technology Policy (OSTP) and the Office of Management and Budget (OMB) work closely with the USGCRP to align the research priorities and funding plans with the Administration's priorities and agency plans. The program recently issued a new strategic plan (see description and link below).

The 2014 Budget proposes $2.7 billion for the USGCRP to support the goals set forth in the program's current strategic plan. These activities can be grouped under the following areas: improve our knowledge of Earth's past and present climate variability and change; improve our understanding of natural and human forces of climate change; improve our capability to model and predict future conditions and impacts; assess the Nation's vulnerability to current and anticipated impacts of climate change; and improve the Nation's ability to respond to climate change by providing climate information and decision support tools that are useful to policymakers and the general public. Reports and general information about the USGCRP are available on the program's website, www.globalchange.gov.

2.1 SELECTED AGENCY HIGHLIGHTS OF THE USGCRP IN THE 2014 BUDGET

- **Understand and Accurately Project Climate Change and its Impacts**. The U.S. Global Change Research Program (USGCRP) integrates Federal research and solutions for climate and global change. The new strategic plan will guide interagency investments in the Budget, including support for a National Climate Assessment of the current science and impacts of climate change. The Budget provides nearly $2.7 billion for USGCRP programs, an increase of $147 million (6 percent) above the FY 2013 enacted level.

    - The Department of Commerce's National Oceanic and Atmospheric Administration (NOAA) is a leading sponsor of oceanic and atmospheric research and is one of the key sponsors of climate science capabilities in the Federal government. The 2014 Budget allocates $371 million for the Department of Commerce's USGCRP efforts, predominantly from NOAA; this represents an increase of $55 million or 17 percent over the FY 2013 enacted level.

    - The National Aeronautics and Space Administration's (NASA) budget includes a sustained investment in climate science, with $1.5 billion proposed for FY 2014. NASA's Earth Science program conducts first-of-a-kind demonstration flights of sensors in air and space in an effort to foster scientific understanding of the Earth system and to improve the ability to forecast climate change and natural disasters. The 2014 Budget supports several research satellites in development, an initiative to monitor changes in polar ice sheets, enhancements to climate models, and NASA contributions to the USGCRP's National Climate Assessment. NASA will continue to develop a replacement to the Orbiting Carbon Observatory (OCO).

- The National Science Foundation (NSF) provides funding for academic basic research across the entire spectrum of the sciences, engineering, and the social sciences. NSF USGCRP support totals $326 million in the 2014 Budget.

- The Department of Energy (DOE) conducts research on climate modeling and predictability that also involves advancing climate and earth system models with improved resolution and uncertainty quantification; DOE also supports long-term atmospheric and terrestrial research experiments. The 2014 Budget allocates $220 million coordinated through USGCRP, with a $7 million increase over FY 2013 dedicated to major field experiments at Arctic, tropics, and oceanic sites. DOE also partners with NSF to support the Community Earth System Model.

- The 2014 Budget provides $72 million for USGCRP programs in the Department of the Interior, an increase of $14 million or 24 percent over the 2013 funding level. Interior's lead science agency, the U.S. Geological Survey (USGS), funds several programs in coordination with other USGCRP agencies to understand the impacts of climate change on natural resources, including the National Climate Change and Wildlife Science Center, which supports a network of Climate Science Centers (CSCs). The CSC supports development of actionable science linked to resource management decisions on climate adaptation.

2.2 LINKAGES TO STRATEGIC PLANS

Interagency Strategic Plans.

- USGCRP 2012-2021 Strategic Plan. This ten-year interagency strategic plan is built around four strategic goals: Advance Science, Inform Decisions, Conduct Sustained Assessments, and Communicate and Educate. In addition to these four goals, the plan emphasizes the importance of national and international partnerships that leverage Federal investments and provide for the widest use of program results. The plan builds on the program's strengths in integrated observations, modeling, and information services for science that serves societal needs. http://downloads.globalchange.gov/strategic-plan/2012/usgcrp-strategic-plan-2012.pdf

- Our Changing Planet. Since 1989 the Global Change Research Program has submitted an annual report to Congress summarizing recent achievements, near term plans, and progress in implementing long term goals. *Our Changing Planet* also provides an overview of recent and near-term expenditures and of requested funding. http://library.globalchange.gov/products/annualreports

Individual Agency Strategic Plans. Excerpts from each participating Agency's strategic plans are provided below along with a weblink to each respective strategic plan.

- Department of Agriculture. Climate change is a central consideration in USDA's strategic planning. Strategic Goal 2 of USDA's Strategic Plan is titled *Ensure our National forests and private working lands are conserved, restored, and made more resilient to climate change, while enhancing our water resources*. USDA also developed a Climate Change Science Plan which presents an overview of the critical questions facing the Department's agencies as they relate to climate change and offers a framework for assessing priorities to ensure consistency with USDA's role in the USGCRP. The objectives of the Climate Change Science Plan include:

- o Restoring and conserving the Nation's forests, farms, ranches, and grasslands.
- o Leading efforts to mitigate and adapt to climate change.
- o Protecting and enhancing America's water resources.
- o Reducing risk from catastrophic wildfire and restore fire to its appropriate place on the landscape.
- o Supporting ecological restoration of our Nation's forests and grasslands and providing research to support improved forest management.
- o http://www.usda.gov/oce/climate_change/science_plan2010/USDA_CCSPlan_120810.pdf

- Department of Commerce. Under its broad goals of generating and communicating new, cutting-edge scientific understanding and promoting economically-sound environmental stewardship and science, the Department of Commerce's Strategic Plan highlights several objectives that will accomplish the following:

- o Advance scientific knowledge and understanding of the Earth's systems, its changing climate, and associated impacts; enhance weather, water, and climate reporting and forecasting; integrate assessments of current and future climate that identify potential impacts; support mitigation and adaptation efforts through sustained, reliable, and timely climate services; and inform the public so that it understands its vulnerabilities to a changing climate and makes informed decisions.
- o http://www.osec.doc.gov/bmi/budget/DOC_Strategic_Plan_022311.pdf

- Department of Energy. DOE's Strategic Plan includes Goal 2: *Maintain a vibrant U.S. effort in science and engineering as a cornerstone of our economic prosperity with clear leadership in strategic areas*; these areas include climate science. The Strategic Plan describes DOE's climate science objective to support *"basic and policy-relevant research underpinning a predictive, systems-level understanding of climate."* To achieve this goal, DOE will:

- o Support fundamental scientific research on climate predictability for improved future projections at the regional spatial scale and with time scales extending from sub-decadal to centennial as part of the U.S. Global Change Research Program and in coordination with the international science community.
- o Provide long-term support to major field research facilities, involving a combination of experimental and modeling activities that focus on atmospheric clouds and aerosols, and terrestrial ecosystems; many of the DOE investments leverage decades of field experience involving sophisticated observational and analytical expertise that has been deployed to sites extending from the Artic to the tropics.
- o Provide long-term support to the comparison, analysis, and diagnosis of all climate models worldwide, in order to enhance US competitiveness in the science of climate predictability.
- o http://energy.gov/sites/prod/files/2011_DOE_Strategic_Plan_.pdf

- Department of Health and Human Services, National Institutes of Health. The FY 2012- 2017 Strategic Plan for the National Institute of Environmental Health Sciences has as its Strategic Goal 5: *Identify and respond to emerging environmental threats to human health, on both a local and global scale.* To achieve this goal NIEHS will:
  - Focus on research needs to help inform policy response in public health situations in which lack of knowledge hampers policymaking, e.g., to improve understanding of the health effects that result from exposures related to climate change.
  - http://www.niehs.nih.gov/about/strategicplan/strategicplan2012_508.pdf

- Department of the Interior. DOI's FY 2011-2016 Strategic Plan contains a strategy to assess and forecast climate change and its effects with its Strategic Goal 2: *Provide Science for Sustainable Resource Use, Protection, and Adaptive Management* and Mission Area 4: *Provide a Scientific Foundation for Decision Making.* This strategy notes that successful adaptation to climate change will depend on access to a variety of options for effective management responses, and describes USGS efforts to:

  - Develop, implement, and test adaptive strategies, reduce risk, and increase the potential for ecological systems to be self-sustaining, resilient, and adaptable to environmental changes.
  - Implement partner-driven science to improve understanding of past and present land use change, develop relevant climate and land use forecasts, and identify lands, resources, and communities that are most vulnerable to adverse impacts of change from the local to global scales.
  - http://www.doi.gov/pfm/upload/DOI_StrategicPlan_FY11-16.pdf

- Department of Transportation. DOT's FY 2012-2016 Strategic Plan includes the Strategic Goal *Advance Environmentally Sustainable Policies and Investments that Reduce Carbon and Other Harmful Emissions from Transportation Sources.* Included in DOT's strategies to achieve this goal are the following:

  - Work through DOT's virtual Center for Climate Change to coordinate climate-related activities, research, and products with the climate experts throughout the Department.
  - Advance aviation climate research to understand the impacts of high-altitude aircraft emissions.
  - Provide technical assistance and incentives to States and Metropolitan Planning Organizations on strategies that reduce GHG emissions.
  - http://www.dot.gov/sites/dot.dev/files/docs/990_355_DOT_StrategicPlan_508lowres.pdf

- Environmental Protection Agency. EPA's FY 2011-2015 Strategic Plan identifies climate change science objectives. Potential impacts of climate change may include increased smog in many regions making it difficult to maintain clean air standards. Climate change may also affect water quality as large volumes of water can overload storm and waste water systems. The Agency's Strategic Plan addresses these challenges in its air and water quality goals.

  - Within EPA's Strategic Goal 1: *Taking action on climate change and improving air quality,* the Strategic Plan identifies an applied research effort to investigate the influence of climate change on clean air, as well as the impacts of emissions from low-carbon fuels in transportation.

- o To achieve EPA's Strategic Goal 2: *Protecting America's water*, EPA will begin to identify actions to respond and adapt to the current and potential impacts of climate change on aquatic resources, including impacts associated with warming temperatures, changes in rainfall amount and intensity, and sea level rise.
- o http://www.epa.gov/planandbudget/strategicplan.html

- National Aeronautics and Space Administration. The 2011 NASA Strategic Plan states as its second Strategic Goal: *Expand scientific understanding of the Earth and the universe in which we live,* and as its Outcome (2.1): *Advance Earth system science to meet the challenges of climate and environmental change.* Within this strategic goal NASA's plan describes several objectives related to climate change science, including:

  - o Improve understanding of and improve the predictive capability for changes in the ozone layer, climate forcing, and air quality associated with changes in atmospheric composition.
  - o Enable improved predictive capability for weather and extreme weather events.
  - o Quantify, understand, and predict changes in Earth's ecosystems and biogeochemical cycles, including the global carbon cycle, land cover, and biodiversity.
  - o Quantify the key reservoirs and fluxes in the global water cycle and assess water cycle change and water quality.
  - o Improve understanding of the roles of the ocean, atmosphere, land and ice in the climate system and improve predictive capability for its future evolution.
  - o Characterize the dynamics of Earth's surface and interior and form the scientific basis for the assessment and mitigation of natural hazards and response to rare and extreme events.
  - o Enable the broad use of Earth system science observations and results in decision-making activities for societal benefits.
  - o http://www.nasa.gov/pdf/516579main_NASA2011StrategicPlan.pdf

- National Science Foundation. NSF's Strategic Plan FY 2011-2016 contains the strategic goal *Innovate for Society* addressing societal needs through research and education, and highlighting the role that new knowledge and creativity play in economic prosperity and society's general welfare. NSF has set Performance Goal (I-1) to *Make investments that lead to results and resources that are useful to society*, and includes a target to:

  - o Support research that underpins long-term solutions to societal challenges such as climate change.
  - o http://www.nsf.gov/news/strategicplan/nsfstrategicplan_2011_2016.pdf

- Smithsonian Institution. SI's FY 2012-2015 Strategic Plan describes its research-related strategic goal to *Advance and synthesize knowledge that contributes to the survival of at-risk ecosystems and species.* The Strategic Plan states an objective to understand how certain environmental stressors including climate change affect the survival of species and the functioning of ecosystems, and includes the following strategies:

- o Enhance the Smithsonian's platforms for long-term research on biodiversity and ecosystems, particularly the Smithsonian Institution Global Earth Observatories (SIGEO).
- o Marshal the Smithsonian's critical mass of biologists and paleontologists, in partnership with experts in other disciplines, to develop understanding of species and ecosystems and find innovative approaches to the complex meta-problems of biodiversity loss, ecosystem degradation, and climate change.
- o http://www.si.edu/content/pdf/about/si_strategic_plan_2010-2015.pdf

- **U.S. Agency for International Development.** In the USAID Policy Framework 2011-2015, a core objective is to reduce climate change impacts and promote low emissions growth. This includes the following research effort:

  - o Finance up to six regional Earth observation hubs to provide over 30 developing countries with better climate change and forecasting data, enabling them to make better decisions in a wide range of areas likely to be affected by climate change.
  - o http://transition.usaid.gov/policy/USAID_PolicyFramework.PDF

Table 2

U.S. Global Change Research Program

Details by Agency/Account

(Budget authority in millions of dollars)[1]

U.S. Global Change Research Program (USGCRP) [1]	FY 2012 Enacted Budget Authority	FY 2013 Enacted Budget Authority	FY 2013 Current Budget Authority[2]	FY 2014 Proposed Budget Authority	Proposed Change in Budget Authority 2013-2014
Department of Agriculture					
Agricultural Research Service	36	36	38	52	+16
National Institute of Food and Agriculture	50	40	40	43	+3
Economic Research Service	2	2	2	2	---
Forest Service – Forest and Rangeland Research	26	25	25	28	+3
National Agricultural Statistics Service	1	1	1	1	---
Natural Resources Conservation Services	1	1	1	1	---
Subtotal – USDA[3]	**116**	**104**	**106**	**126**	**+22**
Department of Commerce					
National Oceanic and Atmospheric Administration – Operations, Research, and Facilities	245	247	233	307	+60
National Oceanic and Atmospheric Administration – Procurement, Acquisition, and Construction	69	64	64	59	-5
National Institute of Standards and Technology (NIST)	5	5	5	5	---
Subtotal – DOC[3]	**319**	**316**	**302**	**371**	**+55**
Department of Energy					
Science – Biological & Environmental Research	**211**	**213**	**209**	**220**	**+7**
Department of Health and Human Services					
Centers for Disease Control and Prevention	6	7	7	7	---
National Institutes of Health	8	8	8	8	---
Subtotal – HHS[3]	**14**	**15**	**14**	**15**	**---**
Department of the Interior					
U.S. Geological Survey – Surveys, Investigations, and Research	**59**	**58**	**55**	**72**	**+14**

U.S. Global Change Research Program (USGCRP) [1]	FY 2012 Enacted Budget Authority	FY 2013 Enacted Budget Authority	FY 2013 Current Budget Authority[2]	FY 2014 Proposed Budget Authority	Proposed Change in Budget Authority 2013-2014
Department of Transportation					
Federal Highway Administration – Federal-Aid Highways[4]	0	0	0	0	---
Federal Aviation Administration – Research, Engineering, and Development	1	1	1	1	---
Federal Transit Administration - Research and University Research Centers[5]	0	0	0	0	---
Subtotal – DOT[3]	**1**	**1**	**1**	**1**	---
Environmental Protection Agency					
Science and Technology	**18**	**19**	**17**	**20**	**+1**
National Aeronautics and Space Administration					
Science	**1,427**	**1,447**	**1,435**	**1,499**	**+52**
National Science Foundation					
Research and Related Activities	**333**	**328**	**316**	**326**	**-2**
Smithsonian Institution					
Salaries and Expenses	**8**	**8**	**8**	**8**	---
U.S. Agency for International Development					
Development Assistance- non-add[6]	*11*	*11*	*11*	*14*	*+3*
Department of State					
Other- non-add[7]	*3*	*3*	*3*	*3*	---
Total[3]	**2,506**	**2,509**	**2,463**	**2,658**	**+149**

Footnotes:

[1] All data supersede numbers released with the 2014 Budget and are current as of June 21, 2013. Budget authority provided in millions of dollars. Any discrepancies are the result of rounding and improved estimates.

[2] Current Budget Authority for FY 2013 throughout this document reflects the amount the program has available for the year calculated as the appropriated amount (as reported in the FY 2013 Enacted column) minus the reductions pursuant to the Budget Control Act of 2011 (P.L. 112-25) sequestration order issued on March 1, 2013, and accounting for any known and applicable reprogrammings, transfers, or other related adjustments. Estimates are current as of June 21, 2013 and are subject to change.

[3] Agency subtotals and table total may not add due to rounding.

[4] The FY 2012 through FY 2014 funding for Federal Highway Administration – Federal Highway Administration – Federal-Aid Highways was less than $500,000.

[5] Federal Transit Administration – Research and University Research Centers is FTA's support for DOT's Center for Climate Change. The FY 2012 through FY 2014 funding amounts for this program are less than $500,000.

[6] USAID funding supports USGCRP and the Climate Change International Assistance effort. In the past, some USAID funding was counted under both categories. These efforts do not add to the USGCRP total.

[7] These efforts do not add to the USGCRP total.

3. CLEAN ENERGY TECHNOLOGIES

Clean Energy Technologies help to reduce, avoid, or sequester greenhouse gas emissions. These programs comprise research, development, and deployment efforts, including a variety of voluntary partnership and grant activities. The activities have the effect of stimulating the development and use of certain energy technologies, including renewable, low-carbon fossil, and nuclear technologies as well as energy efficient technologies, products, and process improvements.

Building on the Administration's progress to make the U.S. the global leader in the clean energy race and protect the environment for generations to come, the 2014 Budget will support American leadership in clean energy. Moving toward a clean energy economy will improve the air we breathe and the water we drink and enhance our energy security by reducing dependence on oil. Clean energy will play a crucial role in slowing global climate change and meeting the President's goals of cutting greenhouse gas emissions in the range of 17 percent below 2005 levels by 2020, and 83 percent by 2050. Just as important, ensuring that the Nation leads the world in the clean energy economy is an economic imperative.

The 2014 Budget proposes approximately $7.9 billion for Clean Energy Technologies. Table 3 provides a breakdown by agency of Clean Energy Technology funding.

Descriptions of some select activities are included below.

3.1 SELECTED AGENCY HIGHLIGHTS OF CLEAN ENERGY TECHNOLOGIES

- **Increased Investment in DOE Climate Change Technology activities.** The Budget proposes $6.2 billion for clean energy technology programs at the Department of Energy, 44 percent more than the 2013 enacted level. The Department's funding supports a wide range of important research, development, and deployment activities on key technologies such as solar, wind, nuclear, and carbon capture and storage. Highlights include:

    - $2.8 billion for the Office of Energy Efficiency and Renewable Energy (EERE) to accelerate research and development, to build on ongoing successes, and to further reduce the costs and increase the use of critical clean energy technologies. Within EERE, the Budget invests $957 million to increase the affordability and convenience of advanced vehicles and domestic renewable fuels and $615 million in innovative projects to make clean, renewable power, such as solar energy and off-shore wind, more easily integrated into the electric grid and as affordable as electricity from conventional sources, without subsidies. It also more than doubles funding to $885 million for energy efficiency and advanced manufacturing activities to help reduce energy use and costs in commercial and residential buildings, in the industrial and business sectors, and in Federal buildings and fleets.

    - $379 million for the Advanced Research Projects Agency - Energy (ARPA-E) to support transformational research in clean energy in areas such as solar energy, energy storage, carbon capture and storage, and advanced biofuels.

3.2 LINKAGES TO STRATEGIC PLANS

Interagency Strategic Plans.

- Blueprint for a Secure Energy Future. In March 2011, the Obama Administration released the *Blueprint for a Secure Energy Future* which outlines the comprehensive national energy policy pursued by the Administration. The *Blueprint* describes strategies across the Federal Government aimed to develop and secure America's energy supplies, provide consumers with choices to reduce costs and save energy and innovate to a clean energy future.
http://www.whitehouse.gov/sites/default/files/blueprint_secure_energy_future.pdf

- Secure Energy Future: Progress Report. In March 2012 accomplishments and achievements that underscore the Administration's commitment to promoting clean energy technologies were described in the *Secure Energy Future: One Year Progress Report.* This report highlights efforts to increase energy independence, set historic fuel economy standards, improve energy efficiency, expand renewable fuel generation, develop advanced alternative fuels and support cutting-edge research. http://www.whitehouse.gov/sites/default/files/email-files/the_blueprint_for_a_secure_energy_future_oneyear_progress_report.pdf

Individual Agency Strategic Plans. Excerpts from each participating Agency's strategic plans are provided below along with a weblink to each respective strategic plan.

- Department of Agriculture. USDA's Strategic Plan for 2010-2015 includes the objective *Lead efforts to mitigate and adapt to climate change* (Strategic Goal 2, Objective 2.2). The plan also includes an objective, *Enhance rural prosperity,* by facilitating sustainable renewable energy development, promoting energy efficiency, and curbing the effects of climate change. (Strategic Goal 1, Objective 1.1). These objectives includes numerous efforts and strategies including:

  o Providing assistance to farmers, ranchers, and forest landowners to implement conservation, nutrient management, and animal management practices that reduce emissions and sequester carbon.
  o Planting and maintaining vegetative cover on marginal farmland and land that has been impacted by fire.
  o Providing assistance in the form of payments, grants, loans and loan guarantees for clean and renewable energy projects and energy efficiency improvements.
  o http://www.ocfo.usda.gov/usdasp/sp2010/sp2010.pdf

- Department of Commerce. The Commerce FY 2011-2016 Strategic Plan includes Objective 5: *Provide the measurement tools and standards to strengthen manufacturing, enabling innovation, and enhancing efficiency*, Objective 6: *Promote and support the advancement of green and blue technologies and industries*, and Objective 16: *Support climate adaptation and mitigation.* Contributing to this objective are the following:

  o Focusing on programs at National Institute of Standards and Technology (NIST) that will develop the measurements, standards, and common framework that are required to

promote sustainable operations and improve energy efficiency in both the construction and manufacturing sectors.
  - o Commerce: (http://www.osec.doc.gov/bmi/budget/DOC_Strategic_Plan_022311.pdf)
  - o NIST: (http://www.nist.gov/director/upload/nist-master-3-year-plan-fy2012-fy2014.pdf)

- Department of Defense. As one of the Government's largest consumers of energy, the Department of Defense is committed to supporting the Administration's efforts in Clean Energy. DOD's Operational Energy Strategy outlines three principles for a stronger force: 1) *Reduce the demand for energy in military operations; 2) Expand and secure the supply of energy to military operations; 3) Build energy security into the future force.* As part of the effort to act on these principles, DOD's Operational Energy Strategy describes goals to:

  - o Reduce energy demand, the most immediate operational energy priority for the Department, by investing in new technologies and equipment.
  - o Expand supply options, both for near-term tactical benefits and long-term operational energy security.
  - o Take energy into account in order to make more informed decisions about the choices and tradeoffs in equipping and employing forces.
  - o http://energy.defense.gov/OES_report_to_congress.pdf

- Department of Energy. DOE's 2011 Strategic Plan lays a framework for utilizing the DOE's capabilities to drive solutions across energy, environmental, climate, and security challenges. It demonstrates strong linkages between clean energy and progress on environmental issues, such as climate change. Shifting to a clean energy economy directly supports the Administration's climate change objective to reduce energy-related greenhouse gas emissions. In particular, Goal 1 of the Strategy: *Catalyze the timely, material, and efficient transformation of the nation's energy system and secure U.S. leadership in clean energy technologies* focuses on activities that support transforming the nation's energy system and building a sustainable and competitive clean energy economy. Targeted outcomes that support this goal include:

  - o DOE and the U.S. Department of Housing and Urban Development working together to enable the cost-effective energy retrofits of a total of 1.1 million housing.
  - o Double renewable energy generation from wind, solar, and geothermal energy sources.
  - o Encourage industry to translate our R&D outputs to market through new contractual vehicles that lower transaction costs and address commercialization barriers.
  - o http://energy.gov/sites/prod/files/2011_DOE_Strategic_Plan_.pdf

- Department of Transportation. DOT's FY 2012-2016 Strategic Plan includes a goal to *Advance environmentally sustainable policies and investments that reduce carbon and other harmful emissions from transportation sources.* Contributing to this goal are strategies to:

  - o Reduce carbon emissions, improve energy efficiency, and reduce dependence on oil, including establishment of fuel economy standards for cars and trucks and research into alternative aircraft fuels.

- o Reduce transportation-related air, water and noise pollution and impacts on ecosystems, including expanding opportunities for shifting freight from less fuel-efficient modes to more fuel-efficient modes.
- o Increase the use of environmentally sustainable practices in the transportation sector, including more environmentally sound construction and operational practices.
- o Reduce pollution from DOT owned or controlled transportation services and facilities, including implementing net-zero-energy building requirements for all new buildings entering the design process in 2020 and thereafter.
- o Promote the deployment of technologies—such as hydrogen fuel cell and diesel-electric hybrid buses—that reduce the energy consumption and greenhouse gas emissions of transit systems.
- o http://www.dot.gov/sites/dot.dev/files/docs/990_355_DOT_StrategicPlan_508lowres.pdf

- **Environmental Protection Agency.** The first goal in EPA's FY 2011-2015 Strategic Plan, *Taking action on climate change and improving air quality,* includes efforts to:

  - o Develop a national system for reporting GHG emissions.
  - o Issue standards to reduce emissions from cars and trucks and non-road sources.
  - o Implement permitting requirements and voluntary programs to promote energy efficiency and encourage design and construction of more efficient processes.
  - o http://www.epa.gov/planandbudget/strategicplan.html

- **National Aeronautics and Space Administration.** The 2011 NASA Strategic Plan states as its Strategic Goal 3: *Create the innovative new space technologies for our exploration, science, and economic future* and Strategic Goal 4: *Advance aeronautics research for societal benefit.* In pursuing these goals NASA is conducting projects that support clean energy technologies as described in several strategic objectives, including:

  - o Develop innovative solutions and technologies to meet future capacity and mobility requirements of the Next Generation Air Transportation System (NextGen).
  - o Develop tools, technologies, and knowledge that enable significantly improved performance and new capabilities for future air vehicles.
  - o Create a pipeline of new innovative concepts and technologies with early-stage Technology Readiness Levels (TRL) for future NASA missions and national needs.
  - o Develop advanced technologies to improve the overall safety of the future air transportation system.
  - o http://www.nasa.gov/pdf/516579main_NASA2011StrategicPlan.pdf

- **National Science Foundation.** NSF's Strategic Plan for 2011-2016 includes a performance goal to *Make investments that lead to results and resources that are useful to society* (Performance Goal I-1). NSF investments underpin long-term solutions to societal challenges such as economic development, climate change, clean energy, and cyber-security:

- Near-term actions include expanding partnerships and collaborations with industry or Government agencies in identifying areas of critical national need and piloting models for investing in priority areas having societal impact.
- Mid-term actions include issuing solicitations and Dear Colleague Letters in areas of critical national need.
- Long-term actions include conducting an impact assessment of the portfolio investments in areas of national need.
- http://www.nsf.gov/news/strategicplan/nsfstrategicplan_2011_2016.pdf

- <u>Nuclear Regulatory Commission</u>. Nuclear is considered a clean energy source, and research at NRC to support its regulatory requirements helps maintain nuclear as part of the clean energy mix going forward. Per the NRC's Strategic Plan, the agency's mission is to protect public health and safety, promote the common defense and security, and protect the environment. NRC's strategic goal on safety: *Ensure adequate protection of public health and safety and the environment* supports clean energy technology by:

  - Implementing focused research programs to anticipate and support resolution of safety issues and address new technologies and conduct research programs to identify and support resolution of longstanding and emergent safety issues.
  - http://www.nrc.gov/reading-rm/doc-collections/nuregs/staff/sr1614/v5/sr1614v5.pdf

- <u>Tennessee Valley Authority</u>. In August 2010, TVA adopted a renewed vision to be one of the nation's leading providers of low-cost and cleaner energy by 2020. TVA's Strategic Plan supports a shift to a cleaner, more efficient and more diverse generating portfolio providing direction to projects, partnerships, and research and development related to:

  - Idling or retiring aging coal units.
  - Increasing generation from nuclear and renewable resources.
  - Promoting energy efficiency and demand response.
  - Exploring and embracing effective new technologies.
  - http://www.tva.com/abouttva/pdf/TVA4-33149_strategic_plan.pdf

Table 3

Clean Energy Technologies

Details by Agency/Account

(Budget authority in millions of dollars)[1]

Clean Energy Technologies[1]	FY 2012 Enacted Budget Authority	FY 2013 Enacted Budget Authority	FY 2013 Current Budget Authority[10]	FY 2014 Proposed Budget Authority	Proposed Change in Budget Authority 2013-2014
Department of Agriculture					
Natural Resources Conservation Service – Conservation Operations	6	6	0	4	-2
Agricultural Research Service – Salaries and Expenses	33	32	32	39	+7
National Institute of Food and Agriculture - Research and Education Activities	31	57	56	51	-5
Forest Service – Commercialization/Renewable Energy	26	23	23	28	+5
Rural Business Cooperative Service – Value Added Producer Grants (Cooperative Development Grants)	1	2	1	1	-1
Rural Business Cooperative Service – Rural Energy Program Account (Rural Energy for America Sec. 9007)	3	3	3	20	+16
Rural Business Cooperative Service –Guaranteed Business and Industry Loans	4	6	5	6	---
Rural Business Cooperative Service – Rural Economic Development Loans[3]	0	0	0	0	---
Economic Research Service[4]	2	2	2	2	---
Office of the Chief Economist - Salaries and Expenses[5]	4	3	3	4	+1
Rural Utilities Service - High Cost Energy Grants[6]	4	4	4	0	-4
2008 Farm Bill, Mandatory Funding					
Rural Business Cooperative Service – Rural Energy Program Account (Rural Energy for America Sec. 9007)	22	0	0	70	+70
National Institute of Food and Agriculture – Biomass Research and Development (Sec. 9008)	40	0	0	26	+26
Farm Service Agency – Biomass Crop Assistance Program	17	0	0	0	---
Farm Service Agency – Commodity Credit Corporation	0	170	161	0	-170
Natural Resources Conservation Service – Farm Security and Rural Investment Programs	16	14	14	14	---
Rural Business Cooperative Service – Energy Assistance Payments (formerly titled Bioenergy Program for Advanced Biofuels (Sec. 9005))	65	0	0	0	---

Clean Energy Technologies[1]	FY 2012 Enacted Budget Authority	FY 2013 Enacted Budget Authority	FY 2013 Current Budget Authority[10]	FY 2014 Proposed Budget Authority	Proposed Change in Budget Authority 2013-2014
Subtotal – USDA discretionary funding	*116*	*138*	*130*	*155*	*+18*
Subtotal – USDA mandatory funding	*160*	*184*	*175*	*110*	*-74*
Subtotal – USDA[7]	**275**	**322**	**305**	**265**	**-57**
Department of Commerce					
National Institute of Standards and Technology (NIST) – Scientific and Technological Research and Services	40	40	40	40	---
National Oceanic and Atmospheric Administration Operations, Research and Facilities	0	0	0	3	+3
Subtotal – Commerce[7]	**40**	**40**	**40**	**43**	**+3**
Department of Defense					
Research, Development, Test and Evaluation, Army	32	29	29	32	+2
Research, Development, Test and Evaluation, Navy	231	186	176	226	+40
Research, Development, Test and Evaluation, Air Force	118	203	190	153	-50
Research, Development, Test and Evaluation, Defense Wide	101	46	42	46	---
Subtotal – DOD[7]	**481**	**465**	**437**	**457**	**-8**
Department of Energy					
Energy Efficiency and Renewable Energy	1,819	1,810	1,719	2,788	+978
Electricity Delivery and Energy Reliability	133	133	126	153	+20
Nuclear Energy	772	765	723	733	-32
Fossil Energy R&D – Carbon Capture and Storage (CCS) and Power Systems	472	446	425	375	-71
Science – Fusion, Sequestration, and Hydrogen	902	924	883	1,067	+143
Energy Transformation Acceleration Fund - Advance Research Projects Agency- Energy (ARPA-E)	275	264	251	379	+114
Bonneville Power Administration Fund[9]	15	17	17	17	---
Race to the Top for Energy Efficiency and Grid Modernization	0	0	0	200	+200
HomeStar	0	0	0	300	+300
Energy Security Trust	0	0	0	200	+200
Subtotal – DOE[7]	**4,388**	**4,359**	**4,144**	**6,212**	**+1,853**

Clean Energy Technologies[1]	FY 2012 Enacted Budget Authority	FY 2013 Enacted Budget Authority	FY 2013 Current Budget Authority[10]	FY 2014 Proposed Budget Authority	Proposed Change in Budget Authority 2013-2014
Department of Transportation					
National Highway Traffic Safety Administration - Operations and Research	10	10	8	11	+1
Research and Innovative Technology Administration – Research and Development	1	1	1	1	---
Federal Aviation Administration - Research, Engineering and Development	21	17	20	18	+1
Federal Aviation Administration - Facilities and Equipment	7	5	4	5	+1
Federal Transit Administration - Research and University Research Centers and Formula and Bus Grants	52	23	22	15	-8
Federal Railroad Association - Railroad Research and Development	1	2	1	3	+1
Subtotal – DOT[7]	**91**	**57**	**56**	**52**	**-5**
Environmental Protection Agency					
Environmental Programs and Management	99	99	95	106	+7
Science and Technology	18	17	16	10	-7
Subtotal – EPA[7]	**117**	**116**	**111**	**115**	**---**
National Aeronautics and Space Administration					
Aeronautics	259	262	255	284	+22
Exploration	9	7	6	9	+1
Space Technology	28	15	15	28	+14
Subtotal – NASA[7]	**296**	**284**	**276**	**321**	**+37**
National Science Foundation					
Research and Related Activities	**341**	**352**	**346**	**372**	**+20**
Nuclear Regulatory Commission					
Salaries and Expenses[8]	**83**	**82**	**57**	**86**	**+4**
Tennessee Valley Authority					
Tennessee Valley Authority Fund[9]	**9**	**11**	**11**	**10**	**-1**
Total[7]	**6,121**	**6,088**	**5,783**	**7,933**	**+1,845**

Clean Energy Technologies[1]	FY 2012 Enacted Budget Authority	FY 2013 Enacted Budget Authority	FY 2013 Current Budget Authority[10]	FY 2014 Proposed Budget Authority	Proposed Change in Budget Authority 2013-2014

Footnotes:

[1] All data supersede numbers released with the 2014 President's Budget and are current as of June 21, 2013. Budget authority provided in millions of dollars. Discrepancies may result from rounding and improved estimates.

[3] Funding for the Rural Business Cooperative Service - Rural Economic Development Loans was less than $500,000 in FY 2012 and FY 2013.

[4] USDA's Economic Research Service has been included in the FCCER, this funding is used to conduct research on the economics of renewable energy.

[5] Office of the Chief Economist includes USDA's Climate Change Program Office and The Office of Energy Policy and New Uses (OEPNU) Research and Development

[6] The Rural Utilities Service - High Cost Energy Grants program has activities to support the creation and use of renewable energy and energy efficiencies.

[7] Agency subtotals and table total may not sum due to rounding.

[8] Nuclear Regulatory Commission funding has been included in the FCCER and reflects funding for nuclear energy research.

[9] Tennessee Valley Authority funding has been added to the FCCER and reflects funding for small modular nuclear reactors research as well as R&D relating to the deployment of nuclear technologies, reduction of greenhouse gas emissions, renewable generation, post-combustion carbon dioxide capture technologies, air quality, energy efficiency and demand response.

[10] Current Budget Authority for FY 2013 throughout this document reflects the amount the program has available for the year calculated as the appropriated amount (as reported in the FY 2013 Enacted column) minus the reductions pursuant to the Budget Control Act of 2011 (P.L. 112-25) sequestration order issued on March 1, 2013, and accounting for any known and applicable reprogrammings, transfers, or other related adjustments. Estimates are current as of June 21, 2013 and are subject to change.

4. INTERNATIONAL ASSISTANCE

The Administration has taken a whole-of-government approach to pursue four broad international climate change financing objectives within its international assistance programs: 1) Demonstrate continued U.S. leadership in forging a global solution to the climate challenge; 2) Help developing countries focus their climate investments strategically over the coming years; 3) Create robust means of measuring, monitoring, and verifying domestic emissions in developing countries; and 4) Reduce vulnerability to climate change. Coordinating and integrating activities from across the U.S. government promotes complementarities that enhance the value of U.S. climate-related financing and increased the likelihood of successfully realizing these four objectives.

Although the Administration's efforts to address climate change are diverse, its bilateral and multilateral international climate change financing is focused on three policy pillars: adaptation, clean energy, and sustainable landscapes. These three policy pillars invest in significant emissions reduction strategies as well as activities that help communities adapt to a changing climate. Key results and indicators to measure progress have been identified for activities in these policy pillars and can be mapped to the Administration's four broad objectives. These activities will strengthen our relationships with other nations, help mitigate the security risk that climate change poses as a threat multiplier in the developing world, support our efforts for a comprehensive, multilateral approach to climate change that involves meaningful actions by all major economies, and create economic opportunities for manufacturers of clean energy technologies.

In its FY 2014 Budget, the Administration is seeking $837 million for core international efforts to combat global climate change, which represents a 5 percent increase from the FY 2013 enacted level. These efforts, conducted by the U.S. Agency for International Development, the U.S. Department of State, and the U.S. Department of the Treasury, will help the most vulnerable countries respond to the growing impacts of climate change, and help forge a global solution to the climate crisis.

The core activities are complemented by an estimated $56 million that is being sought for programs conducted by a range of additional U.S. agencies that address climate change internationally.

In addition to the funding summarized in Table 4, USAID, State, and Treasury will be implementing other programs, such as food security programs or biodiversity programs, in ways that will make a significant contribution to the fight against climate change. Programs focused primarily on non-climate change goals may promote "climate-proofed" development or use adjusted techniques to significantly reduce emissions while promoting other development goals and thereby deliver climate change mitigation and adaptation co-benefits. The Administration estimates the FY 2014 budget authority for these programs to be $216 million.[3] Furthermore, the Administration is enhancing U.S. efforts to address global climate change and promote clean energy technologies in important ways beyond those programs with direct appropriations. Through direct loans, loan guarantees, insurance, and working capital guarantees, U.S. development finance and export credit agencies are increasingly

---

[3] Summary of the "whole of government" U.S. International Climate Change Financing is available at http://www.state.gov/documents/organization/201130.pdf

mobilizing investments in clean energy technologies around the world.[4] These U.S. Government financial products will help American firms, financial institutions, and investors, with their foreign partners, address climate change in developing countries, offering global benefits.

Together, these activities will substantially contribute to the international community's renewed efforts to address climate change, including through the implementation of the Copenhagen Accord, and make clear the Administration's commitment to international leadership in the necessary transition to a low emission economy.

4.1 AGENCY HIGHLIGHTS REGARDING INTERNATIONAL CLIMATE CHANGE ASSISTANCE

- ### U.S. AGENCY FOR INTERNATIONAL DEVELOPMENT (USAID)

USAID is the lead contributor to bilateral assistance, with a focus on capacity building, civil society building, governance programming, and creating the legal and regulatory environments needed to address climate change. USAID will leverage its significant technical expertise to provide leadership in development and implementation of low-carbon strategies, creating policy frameworks for market-based approaches to emissions reduction and energy sector reform, promoting sustainable management of agriculture lands and forests, and mainstreaming adaptation into development activities in countries most at risk. USAID has long-standing relationships with host country governments that will enable it to develop shared priorities and implementation plans. USAID's engagement and expertise in agriculture, biodiversity, health, and other critical climate sensitive sectors provide an opportunity to implement innovative cross-sector climate change programs. Finally, USAID bilateral programs can work in key political and governance areas that multilateral agencies cannot.

- ### DEPARTMENT OF STATE

State takes the lead on diplomatic efforts and deploys financial resources in support of key multilateral and bilateral priorities. State's comparative advantage is promoting effective international solutions, advanced technology strategies, and innovative market approaches through international processes and U.S.-led diplomatic partnerships and initiatives.

- ### DEPARTMENT OF THE TREASURY

The Treasury Department is the primary agency through which the U.S. Government provides contributions through multilateral delivery channels, including the Climate Investment Funds and the Global Environment Facility. Multilateral assistance promotes institutional structures governed jointly by developed and developing countries, which are needed for a coordinated, global response to climate change. Multilateral institutions complement bilateral assistance by

---

[4] Estimates of development finance and export credit agencies' international climate investments are based on an initial review of planned projects, and in some cases the final review of activities, after their implementation, may change the accounting of timing and scale of financing. http://www.state.gov/documents/organization/201130.pdf

leveraging contributions from other donors, making capital investments in infrastructure, providing a range of tailored financial products, and working across a larger number of countries.

The FY 2014 Budget requests $216 million for the Clean Technology Fund (CTF), which aims to close the price gap in developing countries between dirtier conventional technologies and commercially available cleaner alternatives in the power sector, the transportation sector, and in energy efficiency. The CTF focus is on transforming energy use on a sector scale in the "larger emitter" developing countries.

In addition, the FY 2014 Budget requests $68 million for the Strategic Climate Fund in three programs: the Pilot Program for Climate Resilience, the Forest Investment Program, the program for Scaling-up Renewable Energy in Low Income Countries (SREP). The Pilot Program for Climate Resilience will help finance comprehensive efforts to improve the technical capacity of countries to plan for and finance climate adaptation efforts. The Forest Investment Program will support activities informed by national plans to reduce deforestation and will focus on transitioning a small number of developing countries to participate in carbon financing for forest preservation. SREP aims to demonstrate how to put the poorest countries on a pathway that uses renewable energy to expand energy access and stimulate economic growth.

- ## DEPARTMENT OF COMMERCE

The Department of Commerce manages the Renewable Energy and Energy Efficiency Export Initiative (RE4I) through its leadership of the TPCC Working Group on Renewable Energy and Energy Efficiency. The RE4I is a key, sector-specific initiative designed to meet the specific needs of the clean energy sector, while also advancing the President's goal of doubling exports. It includes contributions from eight U.S. Government agencies and is meant to facilitate the deployment of renewable energy and energy efficiency (RE&EE) technologies; better link buyers and sellers of RE&EE products and services; open markets for U.S.-made RE&EE technologies; improve U.S. Government financing for RE&EE exporters; and enhance two-way communication between the U.S. Government and the RE&EE industry. Under the auspices of the RE4I, Commerce has facilitated several improvements to the U.S. Government's trade promotion process in the RE&EE sector, including the development of RE&EE trade policy missions to support existing trade promotion activities by helping to create new markets for U.S. RE&EE companies in countries with nascent policy framework and regulatory systems. This has resulted in highly successful missions to Mexico, Japan, Chile, and Saudi Arabia. Commerce has also developed a first-of-its-kind Renewable Energy Top Prospects Study to help the interagency direct trade promotion activities toward those markets most likely to support U.S. exports.

- ## COMPLEMENTARY AGENCIES

In addition to the core international assistance activities, a number of additional agencies provide technical and in some cases direct support for international efforts to address climate change. Two international agencies, the Millennium Challenge Corporation (MCC) and the U.S. Trade and Development Agency (USTDA), work directly with international partners on projects that may have climate change benefits.

MCC works with its foreign government partners on programs that reduce poverty through sustainable economic growth. In undertaking its poverty reduction programs, MCC seeks to integrate climate change considerations, such as adaptation and reduced emissions, where appropriate. For example, MCC clean energy capital investments may support economic development priorities with ancillary benefits in emissions reductions. MCC agriculture and agricultural infrastructure programs, such as irrigation, may integrate more sustainable use of water resources in those areas at risk of increasing water scarcity in a changing climate.

USTDA has a number of programs that combine support for U.S. exports with a focus on emissions reductions abroad. USTDA provides technical assistance to developing countries on clean energy technologies that U.S. firms provide, organizes visits for foreign entities seeking business opportunities with U.S. firms in the renewable energy sector, and funds studies on future clean energy infrastructure investments.

A number of domestic agencies with significant technical expertise complement the core international assistance activities on climate change through a variety of functions. The Department of Energy and the Environmental Protection Agency provide technical assistance on clean energy investments and environmental regulations undertaken by foreign governments; National Aeronautics and Space Administration and the National Science Foundation provide research and science assistance to the core international assistance agencies that directly supports climate change efforts; and the Forest Service works with USAID on a number of forestry programs that reduce emissions through carbon sequestration.

4.2 LINKAGES TO STRATEGIC PLANS

Interagency Strategic Plans and Planning Documents

- Meeting the Fast Start Commitment – US Climate Finance in Fiscal Year 2012. The Fast Start document released in November 2012 describes the $7.5 billion provided during the three-year fast start finance period from 2010-2012. The three-year fast start finance total consists of more than $4.7 billion of Congressionally-appropriated assistance and more than $2.7 billion from U.S. development finance and export credit agencies.
  http://www.state.gov/documents/organization/201130.pdf

- Fifth Climate Action Report to the UN Framework Convention on Climate Change. The *U.S Climate Action Report 2010* sets out the major actions the U.S. government is taking at the federal level, highlights examples of state and local actions, and outlines U.S. efforts to assist other countries' efforts to address climate change.
  http://www.state.gov/documents/organization/140636.pdf

- Fact Sheet: U.S. Global Development Policy and Global Climate Change Initiative. In September 2010, the President signed a Presidential Policy Directive on Global Development, which provides clear policy guidance to all U.S. Government agencies and enumerates the core objectives, operational model, and the modern architecture needed to implement the policy.
  http://www.whitehouse.gov/sites/default/files/Climate_Fact_Sheet.pdf

Individual Agency Strategic Plans and Planning Documents

- USAID. The *Climate Change and Development Strategy/2012-2016*, released in January 2012, describes USAID's efforts to enable countries to accelerate their transition to climate resilient low emission sustainable economic development. To accomplish this, USAID will pursue three strategic objectives:
  - Accelerate the transition to low-emission development through investments in clean energy and sustainable landscapes;
  - Increase resilience of people, places, and livelihoods through investments in adaptation;
  - Strengthen development outcomes by integrating climate change in Agency programming.
  - http://transition.usaid.gov/our_work/policy_planning_and_learning/documents/GCCS.pdf

- State and USAID. The *Quadrennial Diplomacy and Development Review*, released in 2010, describes the whole-of-government approach used by the Global Climate Change Initiative (GCCI). Key GCCI objectives include:
  - Laying the foundation for low-carbon growth by supporting partner country efforts to advance economic growth while reducing emissions;
  - Accelerating the clean energy revolution through multilateral and bilateral mechanisms and promoting development and deployment of clean energy technologies;
  - Reducing emissions from agricultural and other land use and conserving forests through contributions to Reducing Emissions from Deforestation and Forest Degradation (REDD+).
  - http://www.state.gov/s/dmr/qddr/index.htm

Table 4

International Climate Change Assistance

Details by Agency/Account

(Budget authority in millions of dollars) [1]

International Assistance[1,2]	FY 2012 Enacted Budget Authority	FY 2013 Enacted Budget Authority	FY 2013 Current Budget Authority[20]	FY 2014 Proposed Budget Authority	Proposed Change in Budget Authority 2013-2014
Core Agencies[3]					
Department of State					
Diplomatic and Consular Affairs[4]	0	0	0	0	---
Economic Support Fund	96	95	91	94	-1
International Organizations and Programs	37	37	35	39	+2
Subtotal – State[5]	**133**	**132**	**126**	**133**	**+1**
Department of the Treasury[6]					
Debt Restructuring – Tropical Forestry Conservation	12	12	11	0	-12
Global Environment Facility[7,18]	60	65	63	72	+7
Clean Technology Fund[18]	230	185	175	216	+31
Strategic Climate Fund[8,18]	75	50	47	68	+18
Subtotal – Treasury[5]	**377**	**311**	**296**	**356**	**+44**
U.S. Agency for International Development[4]					
Assistance for Europe, Eurasia, and Central Asia[9]	15	0	0	0	---
Development Assistance[19]	322	322	308	317	-5
Economic Support Fund	12	28	27	32	+4
International Disaster Assistance	0	0	0	0	---
Subtotal – USAID[5]	**348**	**334**	**335**	**349**	**-1**
Subtotal- Core Agencies[5]	*858*	*792*	*757*	*837*	*+45*
Complementary Agencies[10]					
US Department of Energy[11]					
Energy Efficiency and Renewable Energy	9	9	9	9	---
Fossil Energy R&D –Carbon Capture and Storage (CCS)	3	3	3	3	---
Science	1	1	1	1	---

International Assistance[1,2]	FY 2012 Enacted Budget Authority	FY 2013 Enacted Budget Authority	FY 2013 Current Budget Authority[20]	FY 2014 Proposed Budget Authority	Proposed Change in Budget Authority 2013-2014
Subtotal- DOE[5]	13	13	13	13	---
Environmental Protection Agency Environmental Programs and Management[12]	18	18	16	19	---
National Science Foundation Research and Related Activities[14]	6	6	6	3	-3
Department of Agriculture Forest Service-Forest and Rangeland Research[15]	3	3	3	1	-2
National Aeronautics and Space Administration Science[16]	3	3	3	3	---
Millennium Challenge Corporation Millennium Challenge Corporation[17]	41	0	0	0	---
US Trade and Development Agency Trade and Development Agency[21]	16	16	0	18	+1
Subtotal- Complementary Agencies[5]	*100*	*59*	*40*	*56*	*-3*
Total[5]	**958**	**851**	**797**	**893**	**+42**

Footnotes:

[1] This table shows core climate assistance from programs with climate as a primary objective. In addition, indirect climate assistance is provided through programs in other development sectors such as agriculture, water, and health, that do not necessarily have a primary climate objective but nevertheless may provide climate benefits. Those activities have been captured in the U.S. Fast Start Climate Finance Report.

[2] All data supersede numbers released with the 2014 President's Budget and are current as of June 21, 2013. Budget authority provided in millions of dollars. Discrepancies may result from rounding and improved estimates.

[3] Core agencies for the purposes of the Federal Climate Change Expenditure Report are made of the primary climate assistance activities of the Department of State, Department of the Treasury, and US Agency for International Development (USAID). The Federal Climate Change Expenditures Report contained only these core agencies in previous years.

[4] Diplomatic and Consular Affairs continues to support international climate change activities, but because it is not a foreign assistance account, it has been excluded from the international assistance crosscut, beginning in FY 2011.

[5] Agency subtotals and table total may not add due to rounding.

[6]The FY 2012 totals for Treasury climate programming includes a $100 million transfer from the Department of State.

[7]Only 50% of GEF funds are allocated to programs related to climate change and shown here. The full amounts for GEF over respective columns are 2012—120, 2013 enacted—129, 2013 current—125, and 2014 request—144.

[8]The SCF is the second of the multi-donor Climate Investment Funds. It supports three targeted programs: the Pilot Program for Climate Resilience, the Forest Investment Program, and the Program for Scaling-Up Renewable Energy in Low-Income Countries.

[9]In the 2009 Omnibus appropriation, Congress combined Assistance for Eastern Europe and the Baltic States with Assistance for the Independent States of the Former Soviet Union, making a new account called Assistance for Europe, Eurasia, and Central Asia.

[10] The category of Complementary Agencies was first included in the Federal Climate Change Expenditures Report that followed the FY 2011 President's Budget as a means to account for technical and in some cases direct support for international efforts to address climate change.

[11]DOE funding provides global outreach on advanced clean coal technology and CCS for climate change mitigation and energy security in multilateral forums.

[12]EPA activities include Methane to Markets, International Capacity Building, and contribution to the Multilateral Fund to support the Montreal Protocol on Substances that Deplete the Ozone Layer.

[13]ITA funding represents activities under the Asia Pacific Partnership to promote the development and deployment of cleaner and more efficient energy technologies.

[14]NSF funding is for Basic Research to Enable Agriculture Development (BREAD) through the Directorate for Biological Science.

[15]Forest Service activities include assistance to developing countries to establish and maintain sustainable landscape management.

[16]NASA activities include funding for the SERVIR initiative which consists of two web-based regional monitoring networks to provide environmental (land, sea, atmosphere, biota) information and projections to decision makers in Central America/Caribbean and East Africa.

[17] MCC anticipates applying FY 2013 and FY 2014 funds to support compacts with Ghana, Benin, El Salvador, Morocco, Niger, Tanzania, Liberia, and Sierra Leone, which may include funding to support climate change objectives. Because funds will not be committed until signing of a compact and the projects within each compact are still being developed, MCC cannot yet report FY 2013 and FY 2014 funding that will support climate change objectives.

[18] FY 2012 Enacted includes a $100 million ESF transfer from State to Treasury for the Clean Technology Fund ($45 million), Strategic Climate Fund ($25 million), and the Global Environment Facility ($30 million).

[19] FY 2013 ESF funding includes both ESF Base and ESF OCO funds.

[20] Current Budget Authority for FY 2013 throughout this document reflects the amount the program has available for the year calculated as the appropriated amount (as reported in the FY 2013 Enacted column) minus the reductions pursuant to the Budget Control Act of 2011 (P.L. 112-25) sequestration order issued on March 1, 2013, and accounting for any known and applicable reprogrammings, transfers, or other related adjustments. Estimates are current as of June 21, 2013 and are subject to change.

[21] USTDA provides funding for various forms of investment analysis and technical assistance to promote investment opportunities for U.S. companies in developing countries. USTDA has expanded its clean energy project portfolio dramatically over the last few years.

5. ENERGY TAX PROVISIONS THAT MAY REDUCE GREENHOUSE GASES

This report includes existing energy tax provisions and energy payments in lieu of tax provisions which may reduce greenhouse gases. All references to the Code are intended to refer to the Internal Revenue Code of 1986, unless otherwise specified. Summary descriptions of the provisions are provided below and the associated revenue effects are shown in Table 5. A tax expenditure is an exception to baseline provisions of the tax structure that usually results in a reduction in the amount of tax owed. In addition to categories of tax expenditures described in previous Federal climate change expenditures reports, this report contains estimated payments from the Department of the Treasury authorized by Section 1603 of the American Recovery and Reinvestment Act.

All tax expenditure estimates presented here were based upon current tax law enacted as of December 31, 2012. Expired or repealed provisions are not listed if their revenue effects result only from taxpayer activity occurring before fiscal year 2012. Tax expenditure information can also be found in *Analytical Perspectives, Budget of the United States Government*, Fiscal Year 2014, Chapter 16.[5]

Energy production credit – The Code provides a credit for certain electricity produced from wind energy, biomass, geothermal energy, solar energy, small irrigation power, municipal solid waste, qualified hydropower production, or marine and hydrokinetic renewable energy, and sold to an unrelated party.

Energy investment credit – The Code provides credits for investments in solar and geothermal energy property, qualified fuel cell property, qualified microturbine property, geothermal heat pumps, qualified small wind property and combined heat and power property. Owners of renewable power facilities that qualify for the energy production credit may instead elect to take an energy investment credit.

Credit for alternative motor vehicles and refueling property – The Code allows a number of credits for certain types of vehicles and property. These are available for alternative fuel vehicle refueling property, fuel cell vehicles and plug-in electric drive motor vehicles.

Exclusion of utility conservation subsidies – In certain circumstances, public utilities offer rate subsidies to non-business customers who invest in energy conservation measures.

Credit for holding clean renewable energy bonds – The Code provides for the issuance of Clean Renewable Energy Bonds which entitles the bond holder to a Federal income tax credit in lieu of interest. The limit on the volume issued in 2009-2010 is $2.4 billion.

---

[5] Several temporary provisions, including the energy production tax credit, the energy investment credit, the credit for refueling property, the credit for energy efficient improvements to existing homes, the credit for construction of new energy efficient homes, and the credit for energy efficient appliances, were extended or modified under the American Taxpayer Relief Act of 2012. The tax expenditure estimates in this report do not reflect the extension of these tax incentives.

Allowance of deduction for certain energy efficient commercial building property – The Code allows a deduction, per square foot, for certain energy efficient commercial buildings property installed on or in a commercial building.

Credit for construction of new energy efficient homes – The Code allows contractors a tax credit of $2,000 for the construction of a qualified new energy-efficient home with an annual level of heating and cooling energy consumption at least 50 percent below a reference energy standard. The Code also allows a tax credit of $1,000 for the construction of a qualified new energy-efficient manufactured home with an annual level of heating and cooling energy consumption at least 30 percent below a reference energy standard.

Credit for energy efficiency improvements to existing homes – The Code provides an investment tax credit for expenditures made on insulation, exterior windows (including skylights), exterior doors, and metal or asphalt roofs with appropriate pigmented coatings or cooling granules that improve the energy efficiency of a home and meet certain standards. The Code also provides a credit for purchases of advanced main air circulating fans, natural gas, propane, or oil furnaces or hot water boilers, and other qualified energy efficient property.

Credit for residential energy efficient property – The Code provides an investment tax credit for expenditures made on solar electric property, solar hot water heaters, fuel cells, small wind turbines, and geothermal heat pumps for use in a residence.

Credit for energy efficient appliances – The Code provides tax credits for the manufacture of energy efficient dishwashers, clothes washers, and refrigerators. The amount of the tax credit depends on the energy efficiency of the appliance.

Advanced energy property credit – The Code provides a 30 percent investment credit for property used in a qualified advanced energy manufacturing project. The Treasury Department may award up to $2.3 billion in tax credits for qualified investments.

Credit for qualified energy conservation bonds – The Code provides for the issuance of energy conservation bonds which entitle the bond holder to a Federal income tax credit in lieu of interest. The limit on the volume issued in 2009 is $3.2 billion.

Industrial CO2 capture and sequestration tax credit – The Code allows a credit of $20 per metric ton for qualified carbon dioxide captured at a qualified facility and disposed of in secure geological sequestration. The Code also allows a credit of $10 per metric ton of qualified carbon dioxide that is captured at a qualified facility and used as a tertiary injectant in a qualified enhanced oil or natural gas recovery project.

Energy payments in lieu of energy investment credit – Section 1603 of the American Recovery and Reinvestment Tax Act of 2009 (Section 1603) authorizes the Treasury Department to make payments to persons who place in service specified energy property in 2009, 2010, or 2011 or whose construction commenced in 2009, 2010, or 2011. Firms can take an energy payment in lieu of the energy production credit or the energy investment credit.

Table 5
Energy Tax Provisions That May Reduce Greenhouse Gases

(Revenue effect in millions of dollars)

	2012	2013	2014	2015	2016	2017	2018	2014-2018
Energy Production Credit (without coal)[1]	1,452	1,719	1,759	1,719	1,629	1,428	1,088	7,622
Energy Investment Credit[2]	1,040	1,270	1,360	1,670	1,880	1,110	240	6,260
Tax credit for alternative motor vehicles and refueling property[3]	100	180	260	400	610	670	500	2,440
Exclusion of utility conservation subsidies	270	250	250	250	250	250	240	1,240
Credit for holding clean renewable energy bonds	70	70	70	70	70	70	70	350
Allowance of deduction for certain energy efficient commercial building property	70	70	40	20	0	0	-20	40
Credit for construction of new energy efficient homes	70	40	20	0	0	0	0	20
Credit for energy efficiency improvements to existing homes	780	0	0	0	0	0	0	---
Credit for energy efficient appliances	210	300	130	120	100	0	0	350
Credit for residential energy efficient properties[4]	910	1,010	1,140	1,270	1,420	600	0	4,430
Qualified energy conservation bonds	20	30	30	30	30	30	30	150
Industrial CO2 capture and sequestration tax credit	60	60	70	80	110	210	160	630
Tax Provisions Subtotal	*5,052*	*4,999*	*5,129*	*5,629*	*6,099*	*4,368*	*2,308*	*23,532*
Energy Payments in lieu of energy investment credit[2,5]	*5,080*	*8,080*	*4,710*	*2,520*	*1,580*	*330*	*0*	*9,140*
Tax Provisions plus Energy Payments Total	**10,132**	**13,079**	**9,839**	**8,149**	**7,679**	**4,698**	**2,308**	**32,672**

Footnotes:

[1] Estimates of revenue loss from coal provisions have been removed from the tax expenditure estimate in the budget. In previous years, the Expenditures Report cited the New Technology Credit.

[2] In previous years the Energy Investment Credit was contained within the New Technology Credit. The Energy Investment Credit also includes the business installation of fuel cells, which was an independent entry in tables from previous years. These estimates do not exclude microturbine credits which were removed in previous expenditures reports, however the estimates are expected to be too small to affect these figures which are rounded to the nearest $10 million.

[3] In previous reports the tax credit for alternative motor vehicles and refueling property was referred to as the tax credit and deduction for clean-burning vehicles.

[4] In previous years the credit for residential energy efficient property was referred to as the credit for residential purchases/ installations of solar and fuel cells.

[5] Firms can take an energy payment in lieu of the energy investment credit for facilities placed in service in 2009, 2010, or 2011 or whose construction commenced in 2009, 2010, or 2011. The payments are considered outlays and are direct substitutes for the energy tax provisions.

6. CLIMATE ADAPTATION, PREPAREDNESS, AND RESILIENCE

Climate change is a complex, interdisciplinary issue with the potential to affect nearly every sector and level of governmental operations. Across the United States and the world, climate change is already affecting communities, livelihoods, and the environment. To address these challenges and ensure the nation is prepared and resilient to the impacts of climate change, in 2009, the Administration convened the Interagency Climate Change Adaptation Task Force, co-chaired by the Council on Environmental Quality (CEQ), the Office of Science and Technology Policy (OSTP), and the National Oceanic and Atmospheric Administration (NOAA), and including representatives from more than 20 Federal agencies. In addition, on October 5, 2009, President Obama signed an Executive Order directing the Task Force to develop recommendations for how the Federal Government can strengthen policies and programs to better prepare the nation to adapt to the impacts of climate change.

In its 2010 Progress Report, the Task Force called on Federal agencies to demonstrate leadership on climate change adaptation. Rising sea levels, drought, extreme weather events, loss of land and sea ice, and other climate-related impacts threaten communities, ecosystems, and Federal services and assets. The 2010 Task Force Report determined that the Federal Government has a responsibility to safeguard Federal services and resources and to help states, tribes, and communities manage climate-related risks by improving access to climate information, enhancing coordination and capacity, and leading and supporting actions that reduce vulnerability and increase resilience. In response, Federal agencies are taking steps to prepare the nation for the impacts of climate change and are making significant progress. These actions are outlined in agencies' first ever Climate Change Adaptation Plans, which were released in February 2013 as part of the annual Strategic Sustainability Planning Process. These plans outline initiatives to reduce the vulnerability of Federal programs, assets, and investments to the impacts of climate change, such as sea level rise or more frequent or severe extreme weather. Agency adaptation plans highlight actions to plan for and address these impacts in their programs and operations, and protect taxpayer investments.

Agencies are also developing collaborative approaches within the government to build coordinated and comprehensive responses to the impacts of climate change in all sectors. The first of these efforts have focused on building the climate preparedness and resilience of natural resources, including oceans and coasts, wildlife, and water resources. Federal agencies worked with stakeholders to develop a National Action Plan for managing freshwater resources in a changing climate to assure adequate water supplies and protect soil and water quality, human health, property, and aquatic ecosystems. Federal agencies also worked with state, tribal, and local representatives to develop a National Fish, Wildlife and Plants Climate Adaptation Strategy, for safeguarding our nation's species and natural resources (http://www.wildlifeadaptationstrategy.gov). The final strategy was released in March 2013.

There are numerous efforts across the Federal Government for preparing and building resilience to the impacts of climate change on various critical sectors, institutions, and agency mission responsibilities. The President's Climate Action Plan highlights many key efforts to advance climate adaptation, preparedness and resilience. Successful efforts to build resiliency and adaptation often involve integrating climate change considerations into existing agency programs, projects, and activities rather than establishing separate and distinct programs. This creates a challenge when attempting to fully account for all adaptation resources. While the Administration continues to develop methodologies to account for a broader suite of adaptation programs across all critical sectors, this report used the following summary of Department of the Interior activities designed to promote adaptation as an example of one agency's efforts in this area. These Department of the Interior activities also reflect a variety of interagency efforts to address key adaptation challenges that cut across the jurisdictions and missions of individual Federal agencies, and affect fresh water, oceans and coasts, and fish, wildlife and plants.

An early example of agency efforts to promote climate preparedness and resilience is the work of the Department of the Interior to gain effective and broad collaboration to determine the causes and implement changes to reduce climate impacts to lands, waters, natural and cultural resources.

A key component of this initiative is the development of a network of Landscape Conservation Cooperatives (LCCs), which are applied conservation science partnerships that provide scientific and technical support for spatially-explicit conservation goals and for integrated, adaptive management actions at landscape scales. LCCs are composed of and depend on other Federal agencies, tribal, local and State partners, and the public in crafting practical, landscape-level strategies for managing climate change impacts in coordination with the Department's Climate Science Centers (CSCs). The focus of the CSCs includes impacts of climate change on fish, wildlife, and habitats, including wildlife migration patterns, wildfire risk, drought, or invasive species that typically extend beyond the borders of any particular Federal or Tribal land holding.

With resident staff and through connections with partners, LCCs develop, test, implement, and monitor conservation strategies that respond to the dynamic landscape changes resulting from climate change. The LCCs facilitate broad availability of data, modeling, and tools to land managers that allow them to analyze and model trends in species and habitat changes. LCCs also support improved management of water resources, historical and cultural resources, and resources that are needed by Indian Tribes and Alaska Natives.

The FY 2014 President's Budget continues support for cooperative landscape conservation in the face of climate change and other environmental stressors. The National Park Service supports managers with the tools to inventory changes and adapt management practices with a $6 million increase. The Bureau of Indian Affairs provides a $9 million increase to better integrate climate adaptation work on trust land and to support Tribal participation with the CSC and the Landscape Conservation Cooperatives, including the use of traditional ecological knowledge in adaptation management. These funding levels provide the critical science to support Interior's $110 million Climate Change Wildlife Adaptation initiative. The Bureau of Reclamation continues to support cooperative landscape conservation, focusing on climate change implications for water resources management and addressing the Department's Priority Goal for Climate Change through vulnerability assessments, adaptation actions, and development of improved assessment tools through collaboration with CSCs and other climate science entities.

Table 6

Natural Resources Adaptation

(Budget authority in millions of dollars) [1]

Natural Resources Adaptation[1]	FY 2012 Enacted Budget Authority	FY 2013 Enacted Budget Authority	FY 2013 Current Budget Authority[3]	FY 2014 Proposed Budget Authority	Proposed Change in Budget Authority 2013-2014
Department of Interior					
National Park Service – Operation of the National Park Service	3	3	3	9	+6
Fish and Wildlife Service – Resource Management	60	67	64	65	-2
Bureau of Land Management – Management of Lands and Resources	18	18	16	21	+3
Bureau of Indian Affairs – Operation of Indian Programs[2]	0	1	1	10	+9
Bureau of Reclamation – Cooperative Landscape Conservation	7	6	6	5	-1
Total – Natural Resources Adaptation	**88**	**95**	**90**	**110**	**+15**

Footnotes:

[1]All data supersede numbers released with the 2014 President's Budget. Budget Authority provided in millions of dollars. Discrepancies resulted from rounding and improved estimates. Funding in the table does not include USGS climate change adaptation research, which is captured within USGCRP totals.

[2]BIA activities include: assisting Tribes and Alaska Natives with land and resource management, and adaptive management strategies to deal with the effects of climate change they are experiencing or expect to experience; providing climate change funds to Tribes for mitigation and adaptation projects that are deemed high priority; and provide climate change funds for Tribes to actively engage and participate in the Climate Science Centers, LCCs, and the many other climate change implementation projects that require tribal input.

[3] Current Budget Authority for FY 2013 throughout this document reflects the amount the program has available for the year calculated as the appropriated amount (as reported in the FY 2013 Enacted column) minus the reductions pursuant to the Budget Control Act of 2011 (P.L. 112-25) sequestration order issued on March 1, 2013, and accounting for any known and applicable reprogrammings, transfers, or other related adjustments. Estimates are current as of June 21, 2013 and are subject to change.

APPENDIX

ACCOUNTING OF FEDERAL CLIMATE CHANGE EXPENDITURES BY AGENCY

Table 7

Climate Change Expenditures by Agency

Details by Agency/Account
(Budget authority in millions of dollars)[1]

The following is a listing of Federal climate change expenditures by agency and by line item in the President's 2014 Budget Appendix. Budget Appendix line items show account level data and may not reflect sub-account level climate change information. The data in this table may be subsets of an account.

Climate Change Expenditures by Agency	FY 2012 Enacted Budget Authority	FY 2013 Enacted Budget Authority	FY 2013 Current Budget Authority[2]	FY 2014 Proposed Budget Authority	Change in Budget Authority 2013-2014
Department of Agriculture					
Global Change Research Program					
Agricultural Research Service	36	36	38	52	+16
National Institute of Food and Agriculture	50	40	40	43	+3
Economic Research Service	2	2	2	2	---
Forest Service – Forest and Rangeland Research	26	25	25	28	+3
National Agricultural Statistics Service	1	1	1	1	---
Natural Resources Conservation Services	1	1	1	1	---
USDA- GCRP Subtotal	**116**	**104**	**106**	**126**	**22**
Clean Energy Technology					
Natural Resources Conservation Service – Conservation Operations	6	6	0	4	-2
Agricultural Research Service – Salaries and Expenses	33	32	32	39	+7
National Institute of Food and Agriculture – Research and Education Activities	31	57	56	51	-5
Forest Service – Commercialization/Renewable Energy	26	23	23	28	+5
Rural Business Service – Value Added Producer Grants (Cooperative Development Grants)	1	2	1	1	-1
Rural Business Service – Rural Energy for America Program	3	3	3	20	+16
Rural Business Cooperative Service – Guaranteed Business and Industry Loans	4	6	5	6	---
Rural Business Cooperative Service – Rural Economic Development Loans	0	0	0	0	---
Economic Research Service	2	2	2	2	---
Office of the Chief Economist – Salaries and Expenses	4	3	3	4	+1
Rural Utilities Service – High Cost Energy Grants	4	4	4	0	-4

Climate Change Expenditures by Agency	FY 2012 Enacted Budget Authority	FY 2013 Enacted Budget Authority	FY 2013 Current Budget Authority[2]	FY 2014 Proposed Budget Authority	Change in Budget Authority 2013-2014
2008 Farm Bill, Mandatory Funding					
Rural Business Service – Rural Energy for America	22	0	0	70	+70
National Institute of Food and Agriculture – Biomass Research and Development	40	0	0	26	+26
Farm Service Agency – Biomass Crop Assistance Program	17	0	0	0	---
Farm Service Agency – Commodity Credit Corporation	0	170	161	0	-170
Natural Resources Conservation Service – Farm Security and Rural Investment Programs	16	14	14	14	---
Rural Business Service – Bioenergy Program for Advanced Biofuels	65	0	0	0	---
Subtotal - mandatory funding	*160*	*184*	*175*	*110*	*-74*
Subtotal - discretionary funding	*116*	*138*	*130*	*155*	*+18*
USDA- Clean Energy Subtotal	**275**	**322**	**305**	**265**	**-57**
International Assistance					
Forest Service-Forest and Rangeland Research	**3**	**3**	**3**	**1**	**-2**
Total-USDA	**394**	**429**	**414**	**392**	**-37**
Department of Commerce					
Global Change Research Program					
National Oceanic and Atmospheric Administration – Operations, Research, and Facilities	245	247	233	307	+60
National Oceanic and Atmospheric Administration – Procurement, Acquisition, and Construction	69	64	64	59	-5
National Institute of Standards and Technology (NIST)	5	5	5	5	---
DOC- GCRP Subtotal	**319**	**316**	**302**	**371**	**+55**
Clean Energy Technology					
National Institute of Standards and Technology (NIST) – Scientific and Technological Research and Services	40	40	40	40	---
National Oceanic and Atmospheric Administration Operations, Research and Facilities	0	0	0	3	+3
DOC- Clean Energy Subtotal	**40**	**40**	**40**	**43**	**+3**
Total- Department of Commerce	**359**	**356**	**342**	**414**	**+58**

Climate Change Expenditures by Agency	FY 2012 Enacted Budget Authority	FY 2013 Enacted Budget Authority	FY 2013 Current Budget Authority[2]	FY 2014 Proposed Budget Authority	Change in Budget Authority 2013-2014
Department of Defense					
Clean Energy Technology					
Research, Development, Test and Evaluation, Army	32	29	29	32	+2
Research, Development, Test and Evaluation, Navy	231	186	176	226	+40
Research, Development, Test and Evaluation, Air Force	118	203	190	153	-50
Research, Development, Test and Evaluation, Defense Wide[5]	101	46	42	46	---
Total- Department of Defense	**481**	**465**	**437**	**457**	**-8**
Department of Energy					
Global Change Research Program					
Science – Biological & Environmental Research	**211**	**213**	**209**	**220**	**+7**
Clean Energy Technology					
Energy Efficiency and Renewable Energy	1,819	1,810	1,719	2,788	+978
Electricity Delivery and Energy Reliability	133	133	126	153	+20
Nuclear Energy	772	765	723	733	-32
Fossil Energy R&D – Carbon Capture and Storage (CCS) and Power Systems	472	446	425	375	-71
Science – Fusion, Sequestration, and Hydrogen	902	924	883	1,067	+143
Energy Transformation Acceleration Fund – Advance Research Projects Agency- Energy (ARPA-E)	275	264	251	379	+114
Bonneville Power Administration Fund	15	17	17	17	---
Race to the Top for Energy Efficiency and Grid Modernization	0	0	0	200	+200
HomeStar	0	0	0	300	+300
Energy Security Trust	0	0	0	200	+200
DOE- Clean Energy Subtotal	**4,388**	**4,359**	**4,144**	**6,212**	**+1,853**
International Assistance					
Energy Efficiency and Renewable Energy	9	9	9	9	---
Fossil Energy R&D – Carbon Capture and Storage (CCS) and Power Systems	3	3	3	3	---
Science	1	1	1	1	---
DOE- International Assistance Subtotal	**13**	**13**	**13**	**13**	**---**
Adjustments for programs included in multiple categories -- DOE	*-13*	*-13*	*-13*	*-13*	---
Total- DOE	**4,599**	**4,572**	**4,353**	**6,432**	**+1,860**

Climate Change Expenditures by Agency	FY 2012 Enacted Budget Authority	FY 2013 Enacted Budget Authority	FY 2013 Current Budget Authority[2]	FY 2014 Proposed Budget Authority	Change in Budget Authority 2013-2014
Department of Health and Human Services					
Global Change Research Program					
Centers for Disease Control and Prevention	6	7	7	7	---
National Institutes of Health	8	8	8	8	---
HHS- GCRP Subtotal	14	15	14	15	---
Total-HHS	14	15	14	15	---
Department of the Interior					
Global Change Research Program					
U.S. Geological Survey – Surveys, Investigations, and Research	59	58	55	72	+14
Natural Resources Adaptation					
National Park Service – Operation of the National Park Service	3	3	3	9	+6
Fish and Wildlife Service – Resource Management	60	67	64	65	-2
Bureau of Land Management – Management of Lands and Resources	18	18	16	21	+3
Bureau of Indian Affairs – Operation of Indian Programs[2]	0	1	1	10	+9
Bureau of Reclamation – Cooperative Landscape Conservation	7	6	6	5	-1
DOI- Natural Resources Adaptation Subtotal	88	95	90	110	+15
Total-DOI	147	153	145	182	+29
Department of State					
Global Change Research Program					
Other-non-add	*3*	*3*	*3*	*3*	---
International Assistance					
Diplomatic and Consular Affairs	0	0	0	0	---
Economic Support Fund	96	96	91	94	-3
International Organizations and Programs	37	37	35	39	+2
State- International Assistance Subtotal	133	132	126	133	+1
Total-State	133	132	126	133	+1

Climate Change Expenditures by Agency	FY 2012 Enacted Budget Authority	FY 2013 Enacted Budget Authority	FY 2013 Current Budget Authority[2]	FY 2014 Proposed Budget Authority	Change in Budget Authority 2013-2014
Department of Transportation					
Global Change Research Program					
Federal Highway Administration – Federal-Aid Highways	0	0	0	0	---
Federal Aviation Administration – Research, Engineering, and Development	1	1	1	1	---
Federal Transit Administration - Research and University Research Centers	0	0	0	0	---
DOT- GCRP Subtotal	**1**	**1**	**1**	**1**	---
Clean Energy Technology					
National Highway Traffic Safety Administration	10	10	8	11	+1
Research and Innovative Technology Administration – Research and Development	1	1	1	1	---
Federal Aviation Administration - Research, Engineering, and Development	21	17	20	18	+1
Federal Aviation Administration -Facilities and Equipment	7	5	4	5	+1
Federal Transit Administration - Research and University Research Centers and Formula and Bus Grants	52	23	22	15	-8
Federal Railroad Association - Railroad Research and Development	1	2	1	3	+1
DOT- Clean Energy Subtotal	**91**	**57**	**56**	**52**	**-5**
Total-DOT	**92**	**58**	**57**	**53**	**-5**
Department of the Treasury					
International Assistance					
Debt Restructuring – Tropical Forestry Conservation	12	12	11	0	-12
Global Environment Facility	60	65	62	72	+7
Clean Technology Fund	230	185	175	216	+31
Strategic Climate Fund	75	50	47	68	+18
Total-Treasury	**377**	**311**	**296**	**356**	**+44**
Environmental Protection Agency					
Global Change Research Program					
Science and Technology	**18**	**19**	**17**	**20**	**+1**
Clean Energy Technology					
Environmental Programs and Management	99	99	95	106	+7
Science and Technology	18	17	16	10	-7
EPA-Clean Energy Subtotal	**117**	**116**	**111**	**115**	---

Climate Change Expenditures by Agency	FY 2012 Enacted Budget Authority	FY 2013 Enacted Budget Authority	FY 2013 Current Budget Authority[2]	FY 2014 Proposed Budget Authority	Change in Budget Authority 2013-2014
International Assistance					
Environmental Programs and Management	**18**	**18**	**16**	**19**	---
Adjustments for programs included in multiple categories -- EPA	*-9*	*-9*	*-7*	*-9*	---
Total-EPA	**144**	**144**	**137**	**145**	**+1**
Millennium Challenge Corporation					
International Assistance					
Millennium Challenge Corporation	**41**	**0**	**0**	**0**	---
Total-MCC	**41**	**0**	**0**	**0**	---
National Aeronautics and Space Administration					
Global Change Research Program					
Science	**1,390**	**1,444**	**1,428**	**1,493**	**+49**
Clean Energy Technology					
Aeronautics	259	262	255	284	+22
Exploration	9	7	6	9	+1
Space Technology	28	15	15	28	+14
NASA-Clean Energy Subtotal	**296**	**284**	**276**	**321**	**+37**
International Assistance					
Science	**3**	**3**	**3**	**3**	---
Adjustments for programs included in multiple categories -- NASA	*-3*	*-3*	*-3*	*-3*	---
Total-NASA	**1,686**	**1,728**	**1,704**	**1,814**	**+86**
National Science Foundation					
Global Change Research Program					
Research and Related Activities	**333**	**328**	**316**	**326**	**-2**
Clean Energy Technology					
Research and Related Activities	**341**	**352**	**346**	**372**	**+20**
International Assistance					
Research and Related Activities	**6**	**6**	**6**	**3**	**-3**
Total-NSF	**680**	**686**	**668**	**701**	**+15**

Climate Change Expenditures by Agency	FY 2012 Enacted Budget Authority	FY 2013 Enacted Budget Authority	FY 2013 Current Budget Authority[2]	FY 2014 Proposed Budget Authority	Change in Budget Authority 2013-2014
Nuclear Regulatory Commission					
Clean Energy Technology					
Salaries and Expenses	**83**	**82**	**57**	**86**	+4
Total- NRC	**83**	**82**	**57**	**86**	+4
Smithsonian Institution					
Global Change Research Program					
Salaries and Expenses	**8**	**8**	**8**	**8**	---
Total- Smithsonian	**8**	**8**	**8**	**8**	---
Tennessee Valley Authority					
Clean Energy Technology					
Tennessee Valley Authority Fund	**9**	**11**	**11**	**10**	-1
Total- TVA	**9**	**11**	**11**	**10**	-1
US Trade and Development Agency					
International Assistance					
Trade and Development Agency	**16**	**16**	**0**	**18**	+1
Total-TDA	**16**	**16**	**0**	**18**	+1
U.S. Agency for International Development					
Global Change Research Program					
Development Assistance-non-add	11	11	11	14	+3
International Assistance					
Assistance for Europe, Eurasia, and Central Asia	15	0	0	0	---
Development Assistance	322	322	308	317	-5
Economic Support Fund	12	28	27	32	+4
International Disaster Assistance	0	0	0	0	---
USAID-International Assistance Subtotal	**348**	**350**	**334**	**349**	**-1**
Total- USAID	**348**	**350**	**334**	**349**	**-1**

Climate Change Expenditures by Agency	FY 2012 Enacted Budget Authority	FY 2013 Enacted Budget Authority	FY 2013 Current Budget Authority[2]	FY 2014 Proposed Budget Authority	Change in Budget Authority 2013-2014
Total All Agencies[1]	**9,649**	**9,519**	**9,116**	**11,569**	**+2,051**
Energy Tax Provisions That May Reduce Greenhouse Gases	5,052	4,999	4,999	5,129	+130
Energy Payments in lieu of energy investment credit	5,080	8,080	8,080	4,710	-3,370
Total All Agencies + Tax Provisions	**19,781**	**22,598**	**22,195**	**21,408**	**-1,189**

Footnotes:

[1] Totals may not sum due to rounding.

[2] Current Budget Authority for FY 2013 throughout this document reflects the amount the program has available for the year calculated as the appropriated amount (as reported in the FY 2013 Enacted column) minus the reductions pursuant to the Budget Control Act of 2011 (P.L. 112-25) sequestration order issued on March 1, 2013, and accounting for any known and applicable reprogrammings, transfers, or other related adjustments. Estimates are current as of June 21, 2013 and are subject to change.

www.ingramcontent.com/pod-product-compliance
Lightning Source LLC
Chambersburg PA
CBHW081235170526
45165CB00009B/3065